Rourke
Educational Media
rourkeeducationalmedia.com

Animals Have Classes Too!

Birds

Lisa Colozza Cocca

Before, During, and After Reading Activities

Before Reading: Building Background Knowledge and Academic Vocabulary

"Before Reading" strategies activate prior knowledge and set a purpose for reading. Before reading a book, it is important to tap into what your child or students already know about the topic. This will help them develop their vocabulary and increase their reading comprehension.

Questions and activities to build background knowledge:

1. *Look at the cover of the book. What will this book be about?*
2. *What do you already know about the topic?*
3. *Let's study the Table of Contents. What will you learn about in the book's chapters?*
4. *What would you like to learn about this topic? Do you think you might learn about it from this book? Why or why not?*

Building Academic Vocabulary

Building academic vocabulary is critical to understanding subject content.
Assist your child or students to gain meaning of the following vocabulary words.

Content Area Vocabulary

Read the list. What do these words mean?

- *adapted*
- *backbone*
- *classify*
- *common*
- *example*
- *perching*
- *species*
- *wading*

During Reading: Writing Component

"During Reading" strategies help to make connections, monitor understanding, generate questions, and stay focused.

1. *While reading, write in your reading journal any questions you have or anything you do not understand.*
2. *After completing each chapter, write a summary of the chapter in your reading journal.*
3. *While reading, make connections with the text and write them in your reading journal.*
   a) Text to Self – What does this remind me of in my life? What were my feelings when I read this?
   b) Text to Text – What does this remind me of in another book I've read? How is this different from other books I've read?
   c) Text to World – What does this remind me of in the real world? Have I heard about this before? (News, current events, school, etc....)

After Reading: Comprehension and Extension Activity

"After Reading" strategies provide an opportunity to summarize, question, reflect, discuss, and respond to text. After reading the book, work on the following questions with your child or students to check their level of reading comprehension and content mastery.

1. *How do scientists classify birds? (Summarize)*
2. *What can you conclude about birds within a genus as compared to birds within a phylum? (Infer)*
3. *How are hammerheads and storks alike? How are they different? (Asking Questions)*
4. *If you had to make a new way to classify birds, what groups would you sort the birds into? (Text to Self Connection)*

Extension Activity

Cut pictures of birds from old newspapers and magazines. Sort the pictures into groups. Explain to a friend how you decided which birds belong in each group.

Table of Contents

rufous hummingbird

Let's Classify!

There are millions of living things in the world. Scientists **classify** living things to better understand them. They sort them into groups based on the ways things are alike and different.

Animal Groups

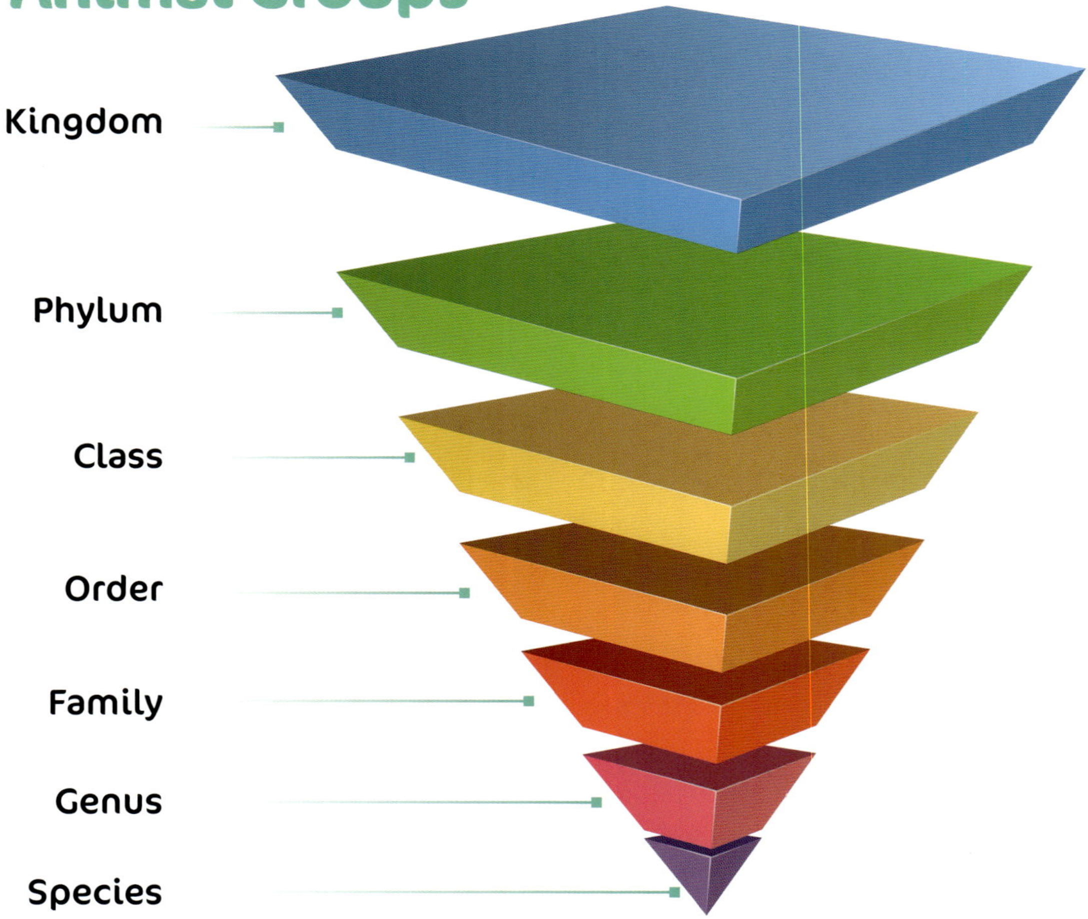

The biggest groups are called kingdoms. Animals make up one kingdom. Members of the animal kingdom eat other living or once-living things. Cows eat grass. Barn owls eat mice. Lobsters eat fish. They all belong to the animal kingdom.

barn owl with rodent

No Plants Allowed!

Plants make their own food, so they are not part of the animal kingdom. They have their own kingdom.

Scientists break kingdoms into smaller groups called phyla. Animals with a **backbone** make up one group, or phylum. Cows and birds are in this group. They have backbones. Lobsters do not have backbones, so they are not in this phylum.

Birds have fewer bones in their body than people have, but they have more bones in their necks. Their 13 to 25 neck bones make it easier for birds to turn their heads to clean their feathers.

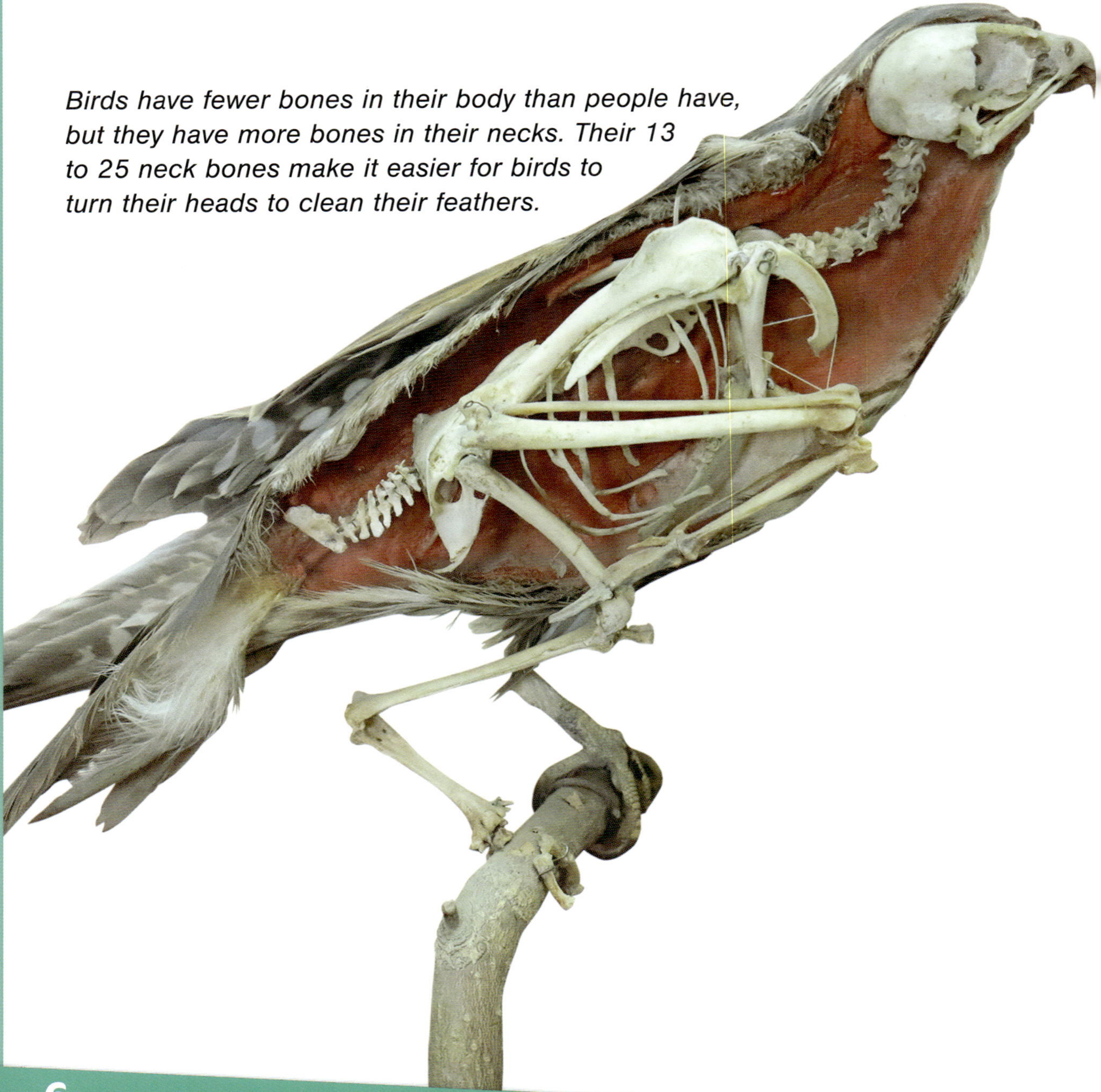

Bird Classes

Classes are smaller groups that make up a phylum. Birds have their own class called Aves. Birds are alike in many ways. All birds have feathers and a beak. They have light bones and strong muscles. They lay eggs with hard shells.

All birds have wings, but not all birds can fly. Blue jays are birds that can fly. Emus are birds, but they cannot fly.

Bats

Only birds have feathers. Bats have wings and fly, but they have no feathers. Bats are not birds.

The Ichthyornis lived millions of years ago. The bird's toothed jaw was more like a dinosaur jaw than a modern bird's.

Many scientists think birds are alike in another way. They have the same ancestors: Ave dinosaurs. The dinosaurs changed over millions of years. They **adapted** to survive in a changing world.

Bird Orders

Scientists break classes into orders. The birds within an order are alike in many ways.

Different Opinions

Scientists don't always agree on the orders. Some sort mostly by how birds are alike. Others look more at differences.

Passeriformes is the largest order of birds. They are **perching** birds. There are thousands of kinds of perching birds. Sparrows, robins, crows, and wrens are in this order. Perching birds have three toes on each foot that point forward. One toe on each foot points backward. Their feet make it easy for them to rest on branches, twigs, and fences.

Songbirds

Perching birds are also known as songbirds.

Some orders are much smaller. Ostriches, emus, and rheas belong to the struthioniformes order. There are only 14 kinds of birds in that order. None of them can fly.

rhea

ostrich

eagle owl

barred owl

great horned owl

barn owl

Owls have their own order called strigiformes. There are more than a hundred kinds of owls.

Families, Genera, and Species

Scientists break orders into smaller groups called families. The ciconiidae and scopidae families are parts of the ciconiiformes order. Medium to tall **wading** birds, such as storks, belong to the ciconiidae family. They have long beaks or bills, long necks, and long, skinny legs.

stork

The hammerhead is also a wading bird. It belongs to the scopidae family. It has short legs, a long beak, and medium-size neck.

hammerhead

Word Endings

The names of bird orders all end with –iformes. The names of bird families all end with –idae.

Families are made up of smaller groups too. Each group is called a genus. Birds in a genus are closely related. A genus is made up of one or more **species**. A bird species is a single kind of bird. Birds in the same species can breed or have babies together.

Northern Cardinal

Kingdom:	Animalia
Phylum:	Chordata
Class:	Aves
Order:	Passeriformes
Family:	Cardinalidae
Genus:	*Cardinalis*
Species:	*Cardinalis cardinalis*

Black-Capped Chickadee

Kingdom:	Animalia
Phylum:	Chordata
Class:	Aves
Order:	Passeriformes
Family:	Paridae
Genus:	*Poecile*
Species:	*Poecile atricapillus*

Scientific Names

Scientific names are in italics or underlined. The genus starts with an uppercase letter, but the species is all lowercase.

A species has a **common** name and a scientific name. Scientific names have two parts. The first part tells what genus to which it belongs. The second part is the name of the species. The scientific name is the same all over the world.

Eastern Bluebird

Kingdom:	Animalia
Phylum:	Chordata
Class:	Aves
Order:	Passeriformes
Family:	Turdidae
Genus:	*Sialia*
Species:	*Sialia sialis*

Great Spotted Woodpecker

Kingdom:	Animalia
Phylum:	Chordata
Class:	Aves
Order:	Piciformes
Family:	Picidae
Genus:	*Dendrocopos*
Species:	*Dendrocopos major*

A common name is often given because of how a bird looks or where it is found. The same name could be given to two different birds living in two different areas. The robin is an **example** of this.

The American robin hunts for worms on the ground. The job of feeding the young is shared by the mother and father.

When settlers arrived in America they saw a small bird with a red breast. It reminded them of the robins in Europe. They called the American bird a robin too.

Scientists studied the birds. They learned both birds eat bugs, worms, and berries. They both live in places with trees such as forests, parks, and backyards.

The birds look different from each other, though. The American robin has a brick-red chest and belly. The European robin has an orange patch on its chest.

Scientists call the American robin *Turdus migratorius*. They call the European robin *Erithacus rubecula*. This tells us the birds are different species. It also tells us each bird belongs to a different genus. They share a common name, but they are not closely related.

The American and European robins belong to the same kingdom, phylum, class, and order.

Scientists continue to learn more about birds. New birds get discovered. The new information means sometimes birds are sorted into a different genus, family, or order.

ACTIVITY

Memory Shakers

People often use short sentences to remember the order of things. You can make up a sentence to help remember the order of the groups of living things. Some examples are:

King **P**aul **C**ooked **O**ur **F**amily's **G**reat **S**oup

Kevin **P**icks **C**arrots **O**ver **F**resh **G**reen **S**pinach

Supplies

paper

pencil

Directions

1. Draw a chart with seven columns. Label the columns from left to right: **K**ingdom, **P**hylum, **C**lass, **O**rder, **F**amily, **G**enus, **S**pecies.
2. Under kingdom, list words that begin with *K*. Under phylum, list words that begin with *P*, and so on.
3. Moving from left to write, pick one word from each column to make a sentence or phrase.

Glossary

adapted (uh-DAPT-ed): changed to better fit or live in an environment

backbone (BAK-bohn): a series of connected bones that run from the skull down through the middle of the back

classify (KLAS-uh-fye): sort into groups of alike things

common (KAH-muhn): used by many people

example (ig-ZAM-puhl): one of a group that shows what the whole group is like

perching (PURCH-ing): sitting or resting above the ground on something easy to fall from, such as a twig

species (SPEE-seez): a group of living things of the same kind with the same name

wading (WAYD-ing): walking through mud, sand, or water

Index

Show What You Know

1. Why do scientists break living things into groups?
2. Describe the feet of a perching bird.
3. Why are the scientific names of birds important?
4. How can you know if a living thing is a bird?
5. How did the American robin get its common name?

Further Reading

Dell, Pamela, *Birds*, Raintree, 2018.

Hughes, Catherine D., *National Geographic Little Kids First Big Book of Birds*, National Geographic Children's Books, 2016.

Lang, Diane, *Fur, Feather, Fin–All of Us Are Kin*, Beach Lane Books, 2018.

About the Author

Lisa Colozza Cocca has enjoyed reading and learning new things for as long as she can remember. She lives in New Jersey by the coast. Her favorite birds to spot are ducks swimming in the pond, seagulls soaring over the ocean, and flamingoes doing anything. You can learn more about Lisa and her work at www.lisacolozzacocca.com.

Meet The Author!
www.meetREMauthors.com

www.rourkeeducationalmedia.com

PHOTO CREDITS: Cover and Title Pg ©KenCanning; Border ©PanuRuangjan; Pg 3 ©kojihirano; Pg 4 ©lvcandy; Pg 5 ©BecCreeper, Pg 6 ©arogant; Pg 7 ©tunart; Pg 8 ©Karine Patry, ©bazilfoto; Pg 9 ©Wiki; Pg 10-11 ©saje; Pg 12 ©JillLang, ©JohnCarnemolla; Pg 13 ©rusm, ©GlobalP, ©mirceax; Pg 14 ©Kaphoto; Pg 15 ©Charles J Sharp (Wiki); Pg 16 ©Dean_Fikar, ©chas53; Pg 17 ©pixhook, ©SteveByland, ©Ocs_12; Pg 18 ©Dean Mitchell; Pg 19 ©JamesBrey; Pg 20 ©ronniechua, ©GlobalPP; Pg 22 ©KREMLL

Edited by: Keli Sipperley
Cover and interior design by: Kathy Walsh

Library of Congress PCN Data

Birds / Lisa Colozza Cocca
(Animals Have Classes Too!)
ISBN 978-1-64369-029-2 (hard cover)
ISBN 978-1-64369-109-1 (soft cover)
ISBN 978-1-64369-176-3 (e-Book)
Library of Congress Control Number: 2018955972

Rourke Educational Media
Printed in the United States of America,
North Mankato, Minnesota